Life in Cellville™

Meet the Cellville Crew

Meet the Power Inside You!

LifeInCellville.com

Life in Cellville™: Meet the Cellville Crew

Meet the Power in You!

Book 1 in the Life in Cellville™ series

Published by GS Creative Works LLC

Gresham, Oregon, USA

Printed in the United States of America

ISBN: 978-1-971524-04-7

Meet the Cellville Crew Song

SCIENCE NOTE

Life in Cellville™ is a fictional, dramatized story inspired by real cell-biology concepts. The characters represent cell structures and organelles and simplify their functions to make complex biological processes fun and easier to understand.

Explore Cellville!
LifeInCellville.com

Explore Life in Cellville™

For a deeper dive into the science behind each organelle, plus free printables, videos, songs, and more, visit:

LifeInCellville.com

Meet The Cellville Crew

Every living thing is made of tiny building blocks called cells. Inside each cell is a bustling little city called Cellville! Just like our bodies have organs (like the heart, lungs, and stomach) that each do a special job, cells have tiny organs called organelles that keep the cell alive and working.

Mindy Membrane
The Gatekeeper!
She lets oxygen in & sends waste out.

Sammy Cytoplasm
The Flow Master!
He helps everything move around & keeps the cell alive.

Nora the Nucleus
The DNA Keeper!
She keeps instructions safe in her DNA scroll.

Ricky & Rudy Ribosomes
The Protein Builders!
They build proteins, the cell's little workers.

Ernie the Rough Endoplasmic Reticulum
The Folding Foreman!
He folds proteins & checks that they are built correctly.

Ellie the Smooth Endoplasmic Reticulum
The Lipid Maker!
She makes membrane-building lipids & helps store calcium.

Gigi Golgi Apparatus
The Package Packer!
She wraps & ships proteins where they need to go.

Mike the Mighty Mitochondrion
The Powerhouse!
He uses energy from food & oxygen to help make ATP.

Perry Peroxisome
The Hazmat Expert!
He breaks down certain fats & helps neutralize harmful chemicals.

Lenny Lysosome
The Recycler!
He recycles old parts & fights germs.

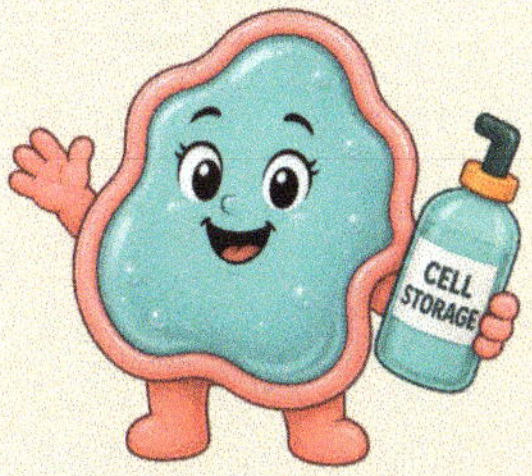

Vicky Vacuole
The Storage Expert!
She keeps materials safely contained inside the cell.

Mindy Membrane guards the gate,
She decides what to let in or wait.
Oxygen slips in, and waste goes out,
She keeps Cellville safe, there's no doubt!

Sammy Cytoplasm fills the space,
A gel-like world, his favorite place.
Not an organelle, but still on the crew,
Everything moves because he knows what to do!

Messages, proteins, waste and more,
Sammy moves them all, shore to shore.
From Nora to Lenny and everyone between,
He's the busiest helper that Cellville has seen!

Nora Nucleus keeps the code,
She stores blueprints in neat rows.
Inside her dome, plans spin and twirl,
Helping build this tiny world!

Ricky and Rudy Ribosomes work side by side,
Tiny but mighty, they beam with pride!
From Nora's blueprints, the twins both read,
Then link small blocks, amino acid beads.

Some join Ernie Rough ER, side by side,
Folding and shipping their proteins with pride.

Some work with Sammy Cytoplasm, floating free,
Building proteins for the cell to see.

Ernie Rough Endoplasmic Reticulum runs the floor,
A maze of belts and pipes galore.
Ricky and Rudy build proteins with care,
Then Ernie folds each one right there.

He checks each fold to make sure it's right,
Then ships them off both day and night.
Down glowing tubes the proteins go,
To Gigi Golgi, next in the flow.

Ellie the Smooth ER stays calm and cool,
Steady and smooth—that's her golden rule.
She's connected to Ernie, side by side,
One winding ER, with two jobs inside.

Her curvy tubes are smooth and bright,
She builds new lipids day and night.
She sends small vesicles, ready to go,
With membrane supplies for Gigi Golgi's flow.

She balances calcium, steady and true,
So muscles can move and nerves can too.
She clears out toxins, keeps danger away,
Keeping Cellville safe, night and day.

Gigi Golgi sorts and packs with cheer,
Her raspberry ribbons stretch far and near.
Proteins from Ernie, lipids from Ellie too,
She tags and ships them - she knows what to do.

Mike the Mighty Mitochondrion
hums all day,
Making bright ATP coins the Cellville way!

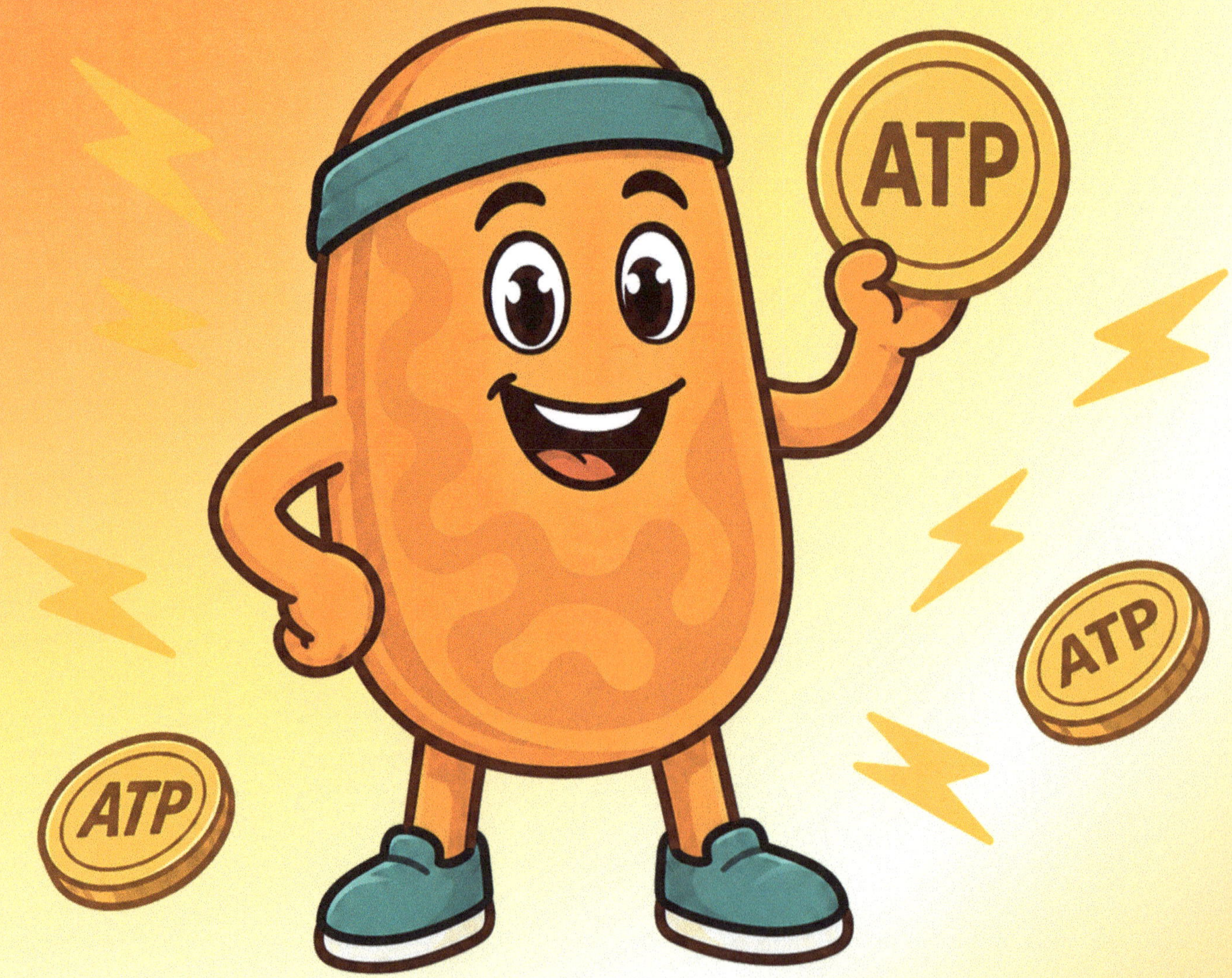

Deep in his coils, where the cristae gleam,
He blends oxygen and fuel in a power stream!
Out burst ATP bolts, glowing and bright,
Powering Cellville day and night!

Perry Peroxisome suits up with care,
Breaking down special fats, he dares!
Working with enzymes, steady and true,
Keeping Cellville safe is what he'll do.

His work makes peroxide, tricky and new,
A chemical spark he must quickly subdue.
He breaks it to oxygen and water so clear,
Cellville's chemical safety engineer!

Vicky Vacuole keeps things contained,
Safe in her membrane, neat and maintained.
From water to waste to nutrients too,
She holds what she needs for the jobs she must do.

Vicky Vacuole brings old parts, worn out or torn,
Then fuses with Lenny to recycle once more.
His enzymes break them into building blocks small,
Ready for Cellville to reuse them all!

Vicky Vacuole traps a germ that doesn't belong,
Then fuses with Lenny, who's ready and strong.
His enzymes break the germ apart,
Helping protect Cellville, they each do their part!

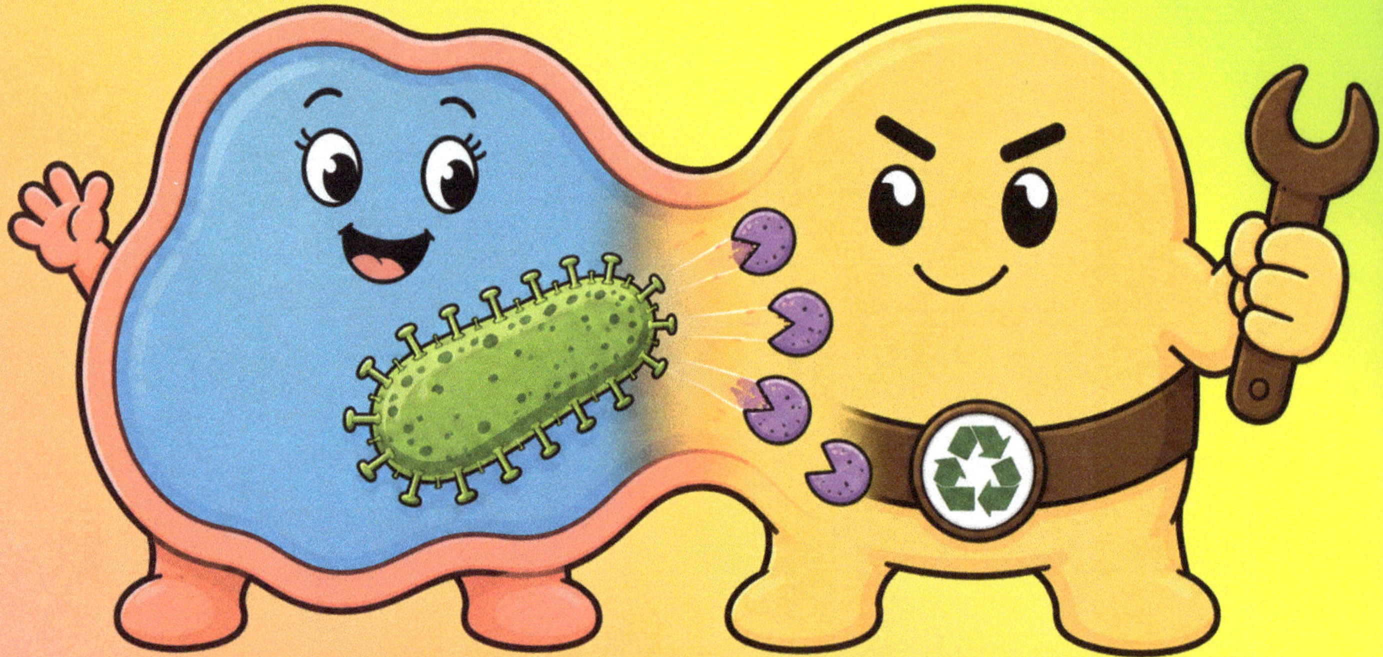

Together they all work, both day and night.
Keeping your body healthy and bright.
From every breath to every chew,
They're all at work, inside of you!

THE END

LifeInCellville.com

Life in Cellville™ WORD POWER!

You met the Crew — now learn their science words!

- Cell — The basic building block of living things.
- Organelle — A tiny structure inside a cell with a special job.
- Cell Membrane — The cell's flexible outer boundary that controls what moves in and out.
- Cytoplasm — The jelly-like material inside the cell where organelles and other cell parts are found.
- Nucleus — Holds and protects the cell's DNA.
- DNA — The genetic instructions cells use to build, grow, and do their jobs.
- Ribosome — Builds proteins.
- Protein — A molecule that helps cells build, repair, and communicate.
- Mitochondrion — Helps turn energy from food into ATP the cell can use.
- ATP — A molecule cells use to power their work.
- Endoplasmic Reticulum (ER) — A membrane network that helps make and process proteins and lipids.
- Golgi Apparatus — Modifies, sorts, packages, and ships cell materials.
- Lysosome — Breaks down and recycles cell materials.
- Vacuole — Stores materials inside the cell.
- Peroxisome — Breaks down certain fats and helps handle harmful chemicals.

Draw a line from the Cellville Crew to their description.

Who’s The Gatekeeper?
She lets oxygen in and sends waste out.

Who’s The Flow Master?
He helps everyone move around and keeps the cell alive.

Who’s The Recycler?
He recycles old parts and fights germs.

Who’s The Folding Foreman?
He folds proteins and checks that each one is built correctly.

Who's The Hazmat Expert?
He breaks down certain fats and helps neutralize harmful chemicals.

Draw a line from the Cellville Crew to their description.

Who's The Lipid Maker?
She makes membrane-building lipids and helps store calcium.

Who's The Storage Expert?
She keeps water, food, and waste safely stored.

Who are The Protein Builders?
They build proteins, the cell's little workers.

Who's The DNA Keeper?
She keeps all the instructions safe in her DNA scroll.

Who's The Powerhouse?
He uses energy from food and oxygen to help make ATP, which powers the cell.

Who's The Package Packer?
She wraps and ships proteins where they need to go.

Meet the Cellville Crew

Meet the Power in You!

Scan to hear Meet the Cellville Crew song!

Chorus

Welcome to Cellville, meet the whole crew,
Every organelle's got a job to do!
Big jobs, small jobs, working through and through,
Meet the Cellville Crew — meet the power in YOU!

Verse 1

Nora Nucleus keeps the blueprints safe and sound,
Mindy Membrane guards the border all around.
Sammy Cytoplasm flows everywhere,
Not an organelle — but he's still working there!

Chorus

Welcome to Cellville, meet the whole crew,
Every organelle's got a job to do!
Big jobs, small jobs, working through and through,
Meet the Cellville Crew — meet the power in YOU!

Verse 2

Ricky & Rudy Ribosomes build proteins down the line,
Ernie Rough Endoplasmic Reticulum folds them up in time.
Ellie Smooth Endoplasmic Reticulum makes lipids fresh and new,
Gigi Golgi packs and ships them — she knows just what to do!

Meet the Cellville Crew

Meet the Power in You!

Scan to hear Meet the Cellville Crew song!

Chorus

Welcome to Cellville, meet the whole crew,
Every organelle's got a job to do!
Big jobs, small jobs, working through and through,
Meet the Cellville Crew — meet the power in YOU!

Verse 3

Mike the Mighty Mitochondrion makes energy for you,
Perry Peroxisome helps with cleanup too.
Lenny Lysosome breaks the old stuff down,
Vicky Vacuole stores materials safe and sound!

Bridge

Even when you're sleeping, tucked in tight,
Cellville never stops, working through the night.
Growing and healing, amazing things they do,
All inside your cells, that's the power in YOU!

Final Chorus

Welcome to Cellville, meet the whole crew,
Every organelle's got a job to do!
Big jobs, small jobs, working through and through,
Meet the Cellville Crew — meet the power in YOU!

Explore Cellville!
LifeInCellville.com

www.ingramcontent.com/pod-product-compliance
Lightning Source LLC
LaVergne TN
LVHW070523061126
851857LV00023B/2136
9781971524047